Is The Universe Really Made Out of Tiny Rubber Bands?

A Kid's Exploration of String Theory

Also available from the MSAC Philosophy Group

Spooky Physics

Darwin's DNA

The Magic of Consciousness

The Gnostic Mystery

When Scholars Study the Sacred

Mystics of India

The Unknowing Sage

In Search of the Perfect Coke

Is the Universe an App?

Adventures in Science

You are Probability

The Mystical

Digital Philosophy

Is The Universe Really Made Out of Tiny Rubber Bands?

A Kid's Exploration of String Theory

By
Shaun-Michael Lane

Mt. San Antonio College
Walnut, California

Is The Universe Really Made Out Of Rubber Bands?

Copyright © 2014 by *Shaun-Michael Diem Lane*

First Print Edition: 2014

ISBN: 978-1-56543-257-4

MSAC Philosophy Group
Mt. San Antonio College
1100 Walnut, California 91789 USA

Website: http://www.neuralsurfer.com

Imprint: *The Runnebohm Library Series*

Publication History: This book was first published in the United States on iBooks as an interactive text for Apple's iPad. It quickly became the biggest selling book on string theory on iTunes, eventually becoming number one in its category for over two straight years.

Manufactured in the United States of America

Acknowledgments

I would like to express my thanks to all those who helped me in doing this book. First and foremost, I want to say thank you to my wonderful 5th grade teacher Mr. Hammond. Second, I want to say thanks to my mom and dad who answered my questions and helped guide me. Third, I want to acknowledge the valuable insights I learned from reading and watching Professor Brian Greene from Columbia University. His books and lectures are very informative and clear. I sent him a copy of my book and the film I made based upon it and he kindly wrote back saying that he enjoyed it very much and encouraged me to study math and physics as I grow up. Finally, I want to say thanks to my little brother Kelly for playing Minecraft with me when I needed to take a break.

--Shaun-Michael Lane, *Huntington Beach*

Is the Universe Really Made of Tiny Rubber Bands?

A Kid's Exploration of String Theory

Hi, my name is Shaun. However, my parents' pet name for me is Plato.

I have always wondered about the origin of matter. Once I took an old golf ball and broke it apart. I was surprised to see that there were tightly woven rubber bands inside.

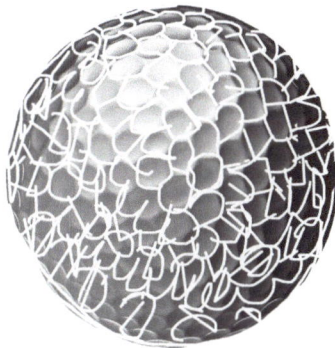

This got me to thinking about what made up rubber bands. So I went to my father and mother and they suggested that I look into the subject of physics, but more specifically into the subject of quantum mechanics, which is the study of how very small things behave.

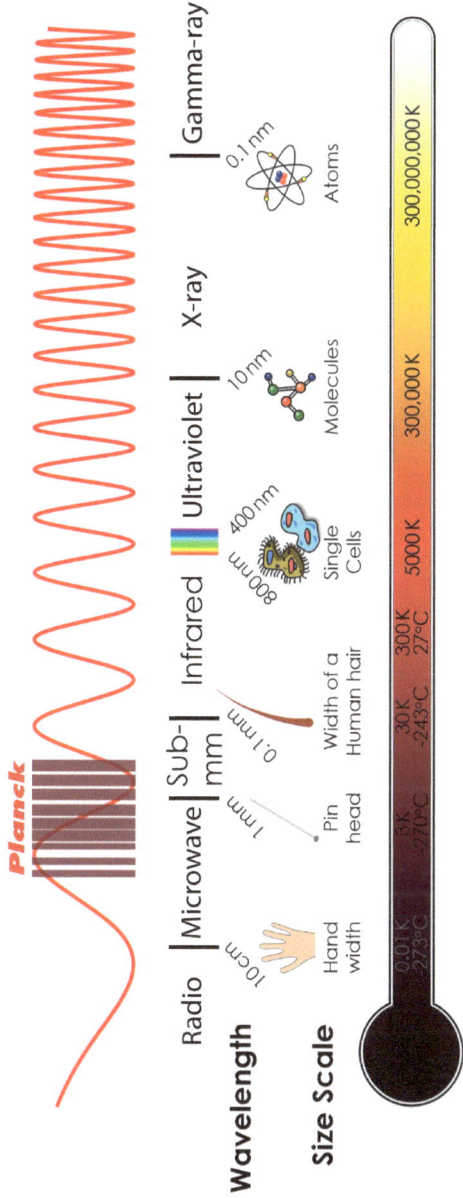

Wavelength

Radio | Microwave | Sub-mm | Infrared | Ultraviolet | X-ray | Gamma-ray

10cm | 1 mm | 0.1 mm | 800 nm / 400 nm | 10 nm | 0.1 nm

Planck

Size Scale

Hand width | Pin head | Width of a Human hair | Single Cells | Molecules | Atoms

0.01 K
-273°C | 3 K
-270°C | 30 K
-243°C | 300 K
27°C | 5000 K | 300,000 K | 300,000,000 K

5

Richard Feynman, a very famous scientist who worked on developing the first atom bomb, was once asked if he could try to explain all of physics in one simple sentence. His reply was pretty funny, but quite profound: "Things are made of littler things that jiggle."

Richard Feynman won the Nobel Prize in Physics for his work on Quantum Electrodynamics.

So if this was true," I thought to myself, "then I could unlock the secret of matter by breaking things down to their smaller parts and seeing how they moved around."

When I took apart a rubber band I found out they were made of even smaller lines of rubber. But when I went and pulled those apart all I got were little pieces.

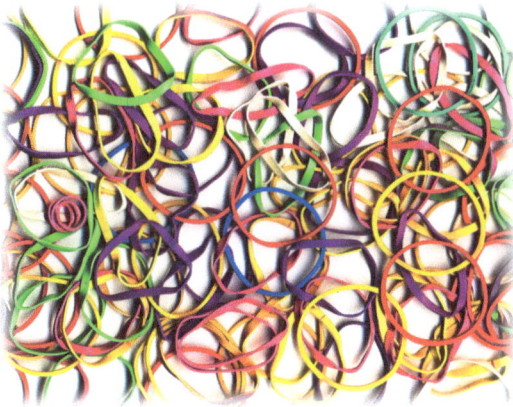

9

I couldn't go any farther with my bare hands so I got my microscope out and when I got the piece positioned just right I could see that it was made of smaller thread like material.

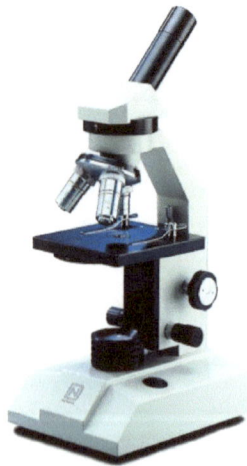

"What is that stuff?"
I pondered.

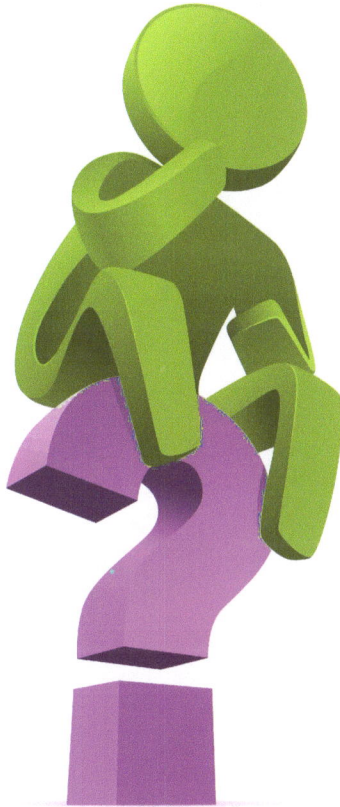

What to do?

I then realized I had to learn from the great physicists in the past who have explored this subject in depth.

This is when I learned about the early Greek philosopher, Democritus, who suggested that all things are made up of little tiny balls which he called atoms which he thought were <u>un</u>cuttable.

This turned out later to be wrong since atoms can be broken apart, but the word atoms became a popular way of talking about really tiny things.

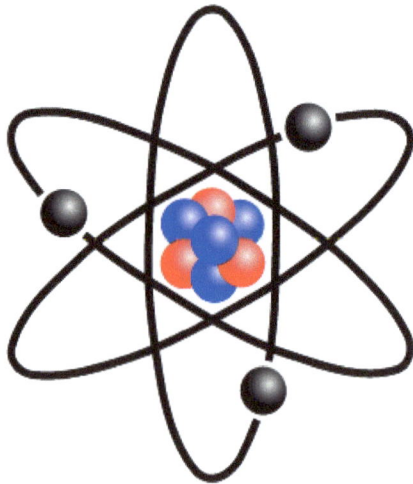

In the 19th and 20th century, physicists realized that atoms were comprised of mostly empty space. But in their core was a seed like thing called a nucleus which houses smaller bits of matter called protons and neutrons.

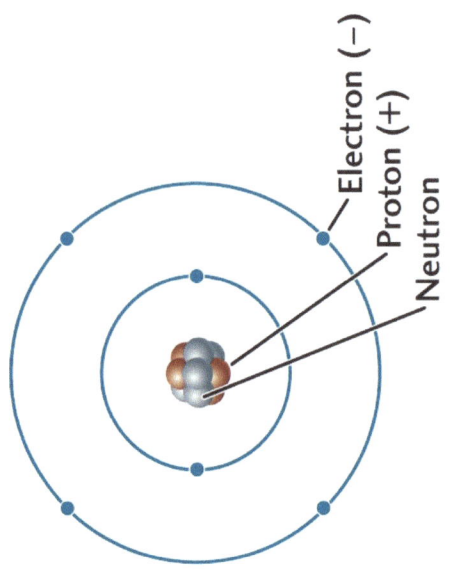

Electron (−)
Proton (+)
Neutron

Carbon atom

Electron cloud

Nucleus

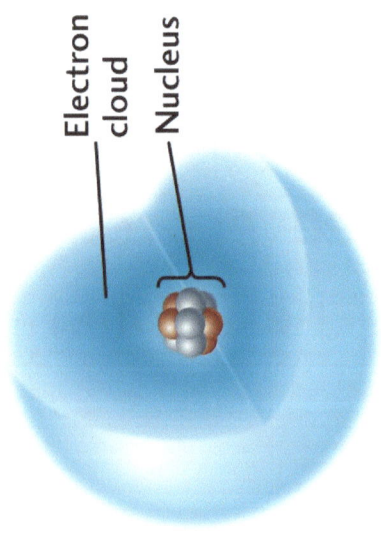

16

And darting around that nucleus seed was a superfast particle called an electron.

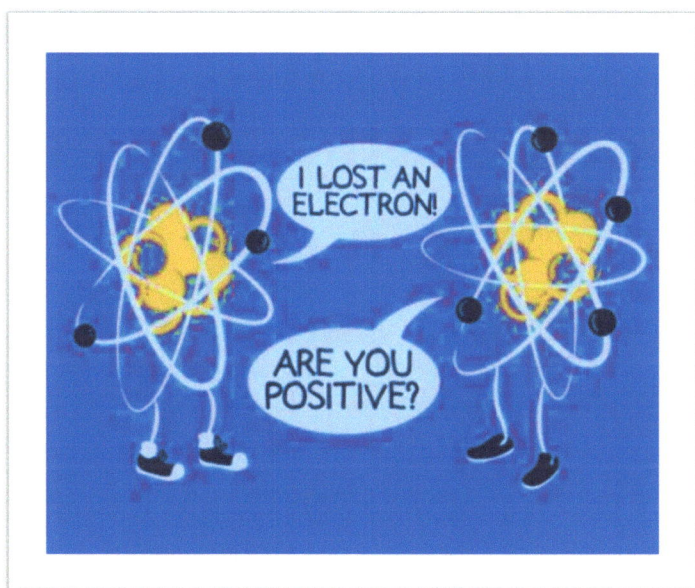

I found it intriguing that the number of protons and electrons in an atom are recognized as different chemical elements such as hydrogen (the simplest atom with only one proton and one electron) and uranium (an atom with 92 protons and 92 electrons).

Periodic Table of the Elements

Legend	
Alkali Metal	
Alkaline Earth	
Transition Metal	
Basic Metal	
Semimetals	
Nonmetals	
Halogens	
Noble Gas	
Lanthanides	
Actinide	

These different atomic numbers explain why some things are solid and dense, such as gold, and other things are airy and flighty such as when you put helium in a balloon.

Everything we see around us is just the mixing of different types of atoms.

Yet, this got me to look further into the heart of atoms. What is inside protons and neutrons? What is light?

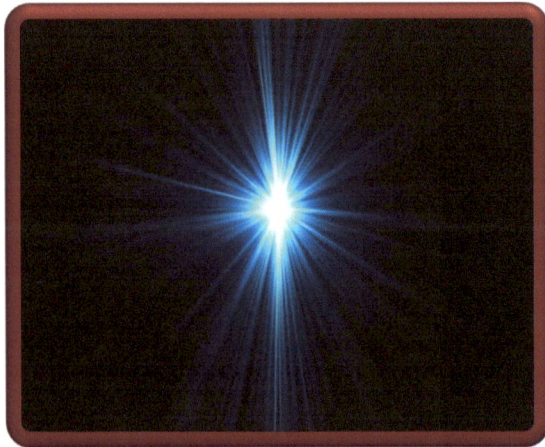

Scientists such as Max Planck, Albert Einstein, Niels Bohr, Max Born, Erwin Schrodinger, Paul Dirac, Wolfgang Pauli, and Werner Heisenberg developed a new way of understanding physics which became popularly known as quantum mechanics.

23

SOLVAY CONFERENCE 1927

A. PICARD   E. HENRIOT   P. EHRENFEST   Ed. HERSEN   Th. DE DONDER   E. SCHRÖDINGER   E. VERSCHAFFELT   W.PAULI   W. HEISENBERG   R.H FOWLER   L. BRILLOUIN

P. DEBYE   M. KNUDSEN   W.L. BRAGG   H.A. KRAMERS   P.A.M. DIRAC   A.H. COMPTON   L. de BROGLIE   M. BORN   N. BOHR

I. LANGMUIR   M. P. ANCK   Mme CURIE   H.A.LORENTZ   A. EINSTEIN   P. LANGEVIN   Ch.E. GUYE   C.T.R. WILSON   O W. RICHARDSON

Absents : Sir W.H. BRAGG, H. DESLANDRES et E. VAN AUBEL

24

Just as an auto mechanic tries to understand what is wrong with your car by opening up the hood and looking closely at the car's engine, the quantum mechanic looks under the hood of the atom to see how the tiny bits of matter are working together.

Physicists were shocked to realize that matter at very small scales behaves in completely unexpected ways.

Indeed the more they tried to figure out the exact location and speed of an electron, they noticed that their very act of trying to measure both interfered with the electron. Thus, scientists couldn't know both precisely.

THE UNCERTAINTY PRINCIPLE

Said one electron to another, "Where are you?"
"I'm certain that I'm over here!" said the other.
"How fast can you get here?"
"That," said the electron, "I couldn't tell you."

Imagine that you are rolling a pair of dice outside in the dark and you couldn't see what numbers you rolled.

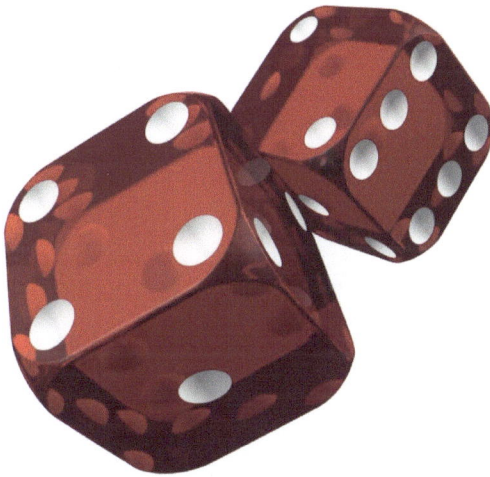

However, when you put your flashlight on the dice, the light itself changed the roll of the dice so that you couldn't know what your original throw was.

Your very act of shining light on the dice literally altered your throw.

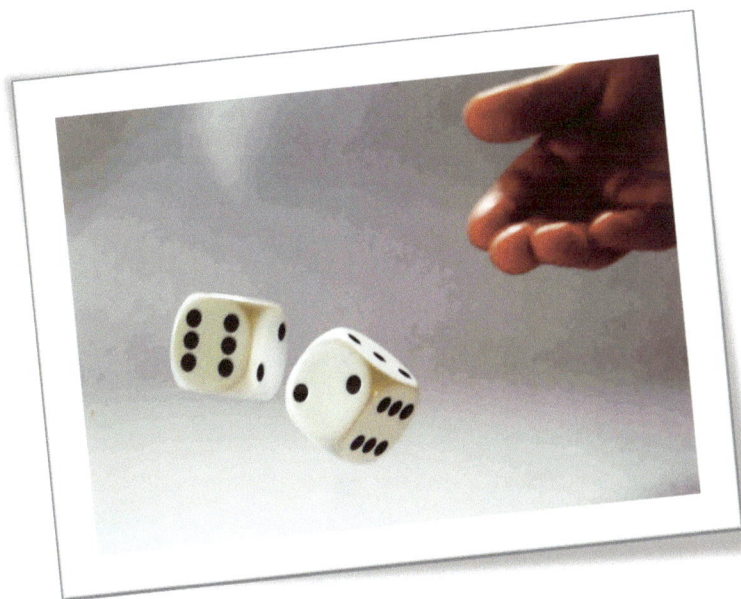

This is very similar to what happens when physicists try to figure out what electrons are really doing only to realize that their very act of observing changed what they observed.

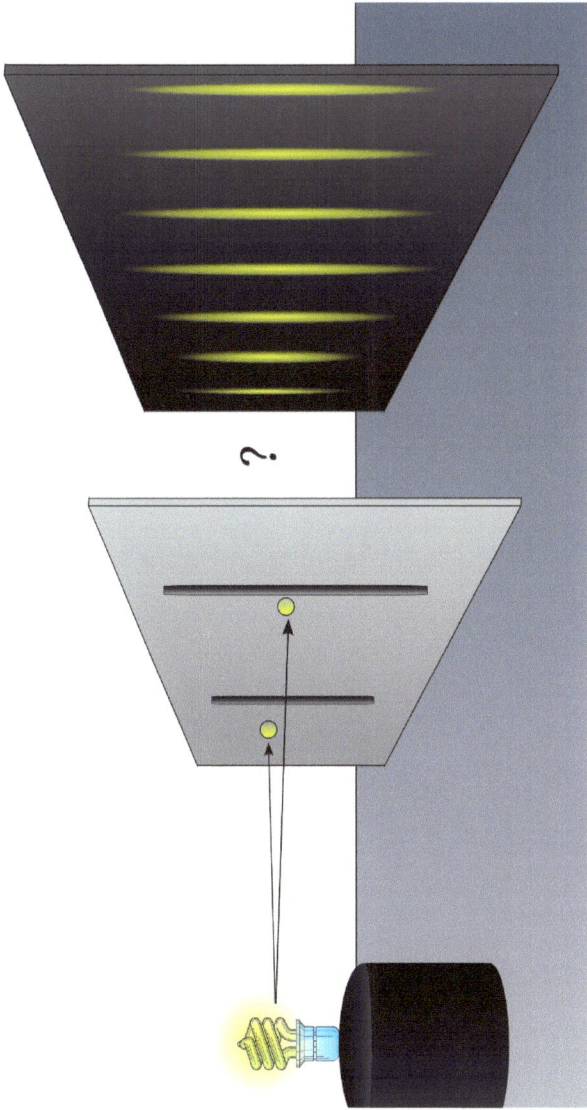

This understanding caused a revolution in physics and started them on a new course where chance and probability played an important part in studying the secrets of nature.

They realized
that there were four
fundamental forces
in the universe.

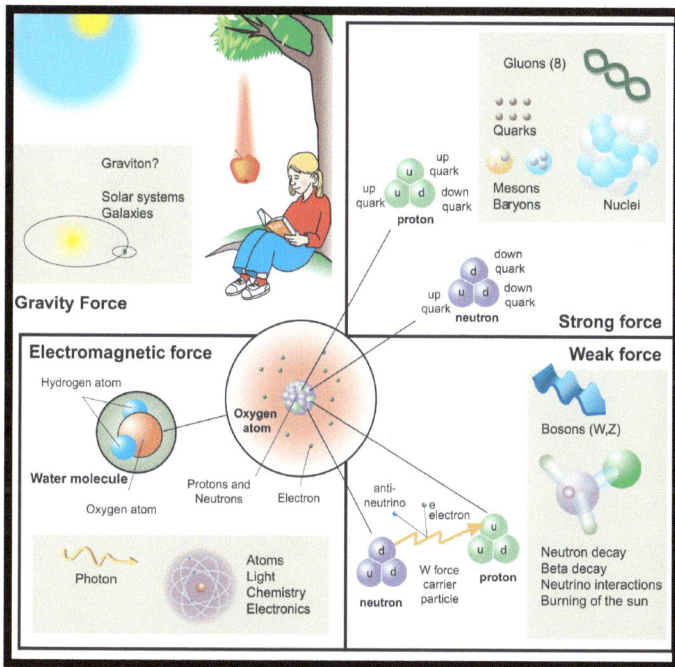

1. Electromagnetism (which is where we get light and electricity and our food from).

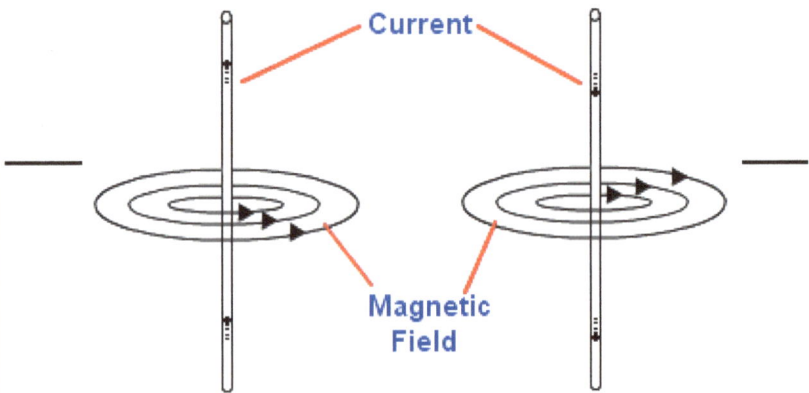

Current

Magnetic Field

2. Strong nuclear force (this keeps protons and neutrons bonded together for the most part).

3. Weak nuclear force (which explains why there is radioactive decay within atoms).

RADIATIVE NEUTRON BETA DECAY

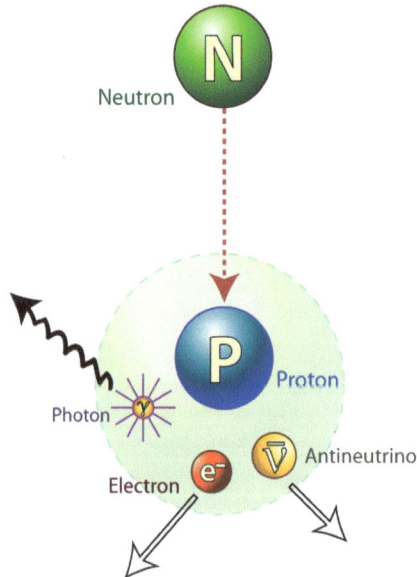

Neutron

Proton

Photon

Electron

Antineutrino

4. Gravity (which is the universal attractive force which explains why the earth orbits the sun and why I cannot jump very high when I play basketball).

But scientists wondered if all four of these forces were the result of one very tiny super process. They called this quest a G.U.T., which stands for Grand Unified Theory or a T.O.E., a Theory of Everything.

GUT

This made me laugh thinking that the hidden mystery of the universe was in a GUT or a TOE.

THEORY OF EVERYTHING

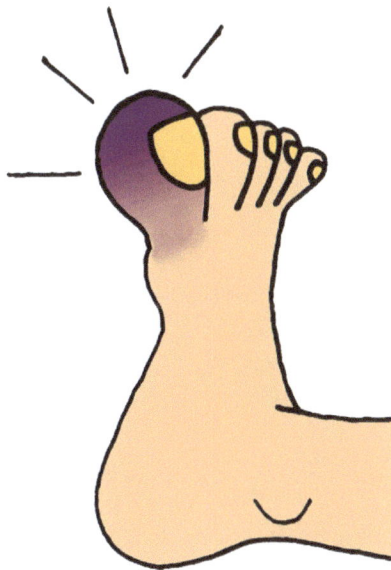

Scientists realized that to find out the hidden parts of atoms they had to investigate even smaller bits. In order to do this they had to force protons to break apart so they could see what they contained.

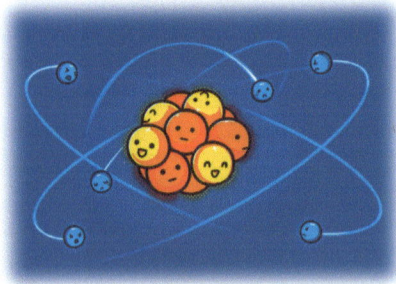

Generation of Particle Beams Using High-intensity Proton Beams

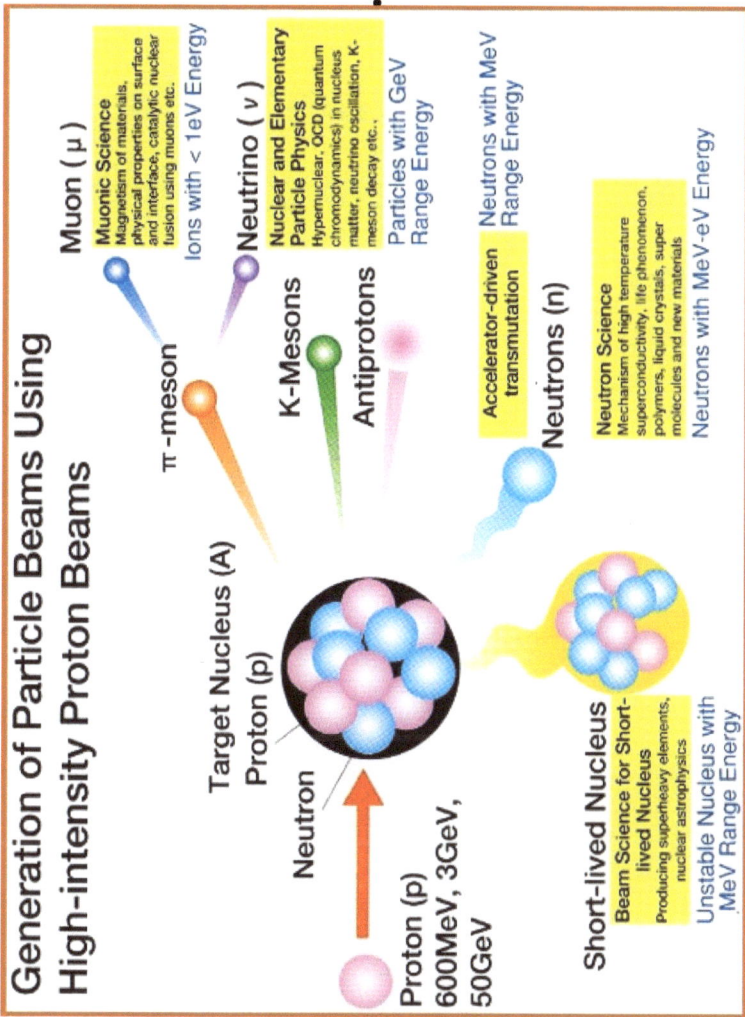

Proton (p)
600MeV, 3GeV, 50GeV

Target Nucleus (A)
Proton (p)

Neutron

Muon (μ)

Muonic Science
Magnetism of materials, physical properties on surface and interface, catalytic nuclear fusion using muons etc.

Ions with < 1eV Energy

π -meson

Neutrino (ν)

Nuclear and Elementary Particle Physics
Hypernuclear, QCD (quantum chromodynamics) in nucleus matter, neutrino oscillation, K-meson decay etc..

Particles with GeV Range Energy

K-Mesons

Antiprotons

Accelerator-driven transmutation

Neutrons with MeV Range Energy

Neutrons (n)

Neutron Science
Mechanism of high temperature superconductivity, life phenomenon, polymers, liquid crystals, super molecules and new materials

Neutrons with MeV-eV Energy

Short-lived Nucleus

Beam Science for Short-lived Nucleus
Producing superheavy elements, nuclear astrophysics

Unstable Nucleus with MeV Range Energy

43

So scientists built big colliders where they had protons zipping around near the speed of light in large circular tracks until they hit each other and broke apart.

This reminded me of when I would occasionally take a shell or a rock at the beach and try to open it apart and see what is inside of it.

The secret of the universe is in the tiniest bits of matter.

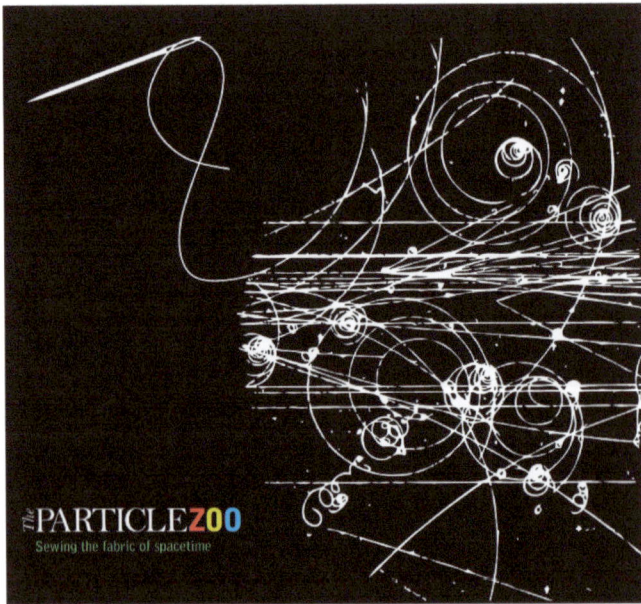

THE PARTICLE ZOO
Sewing the fabric of spacetime

In the 1980s a new theory was developed called String Theory which essentially claims that everything in the universe is made out of very small loops of matter, much, much smaller than anything we can see in atoms.

47

These strings vibrate in various ways and thus cause various forms of matter to be created.

All the elementary particles that make up our universe are like musical notes that are caused by infinitesimally small loops of guitar like string (sometimes open and sometimes closed).

Strike a different vibration of one of these strings and a different physical property emerges. These strings, however, are so small that nobody can see them, not even with the world's most powerful microscope.

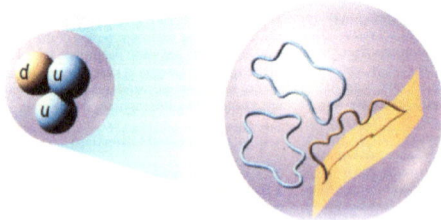

This is why scientists built a Large Hadron Collider in Switzerland which is pretty deep underground so they could smash protons against each other and try to recreate the initial conditions shortly after the Big Bang.

BANG!

The Big Bang is when our universe started 13.8 billion years ago.

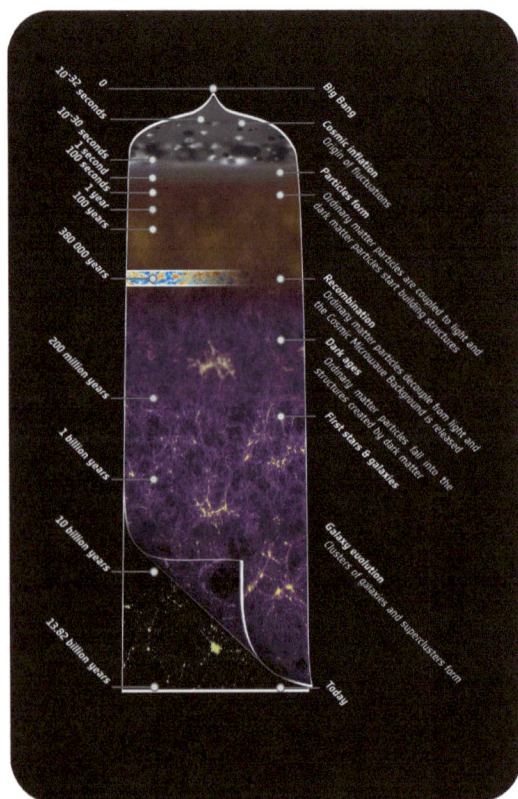

Everything back then was collapsed into a tiny little seed, including all the stars and the planets. This single point was much smaller than a penny but it was amazingly heavy with all that packed in energy and matter. So when it exploded it was called a "Big Bang."

In this way scientists realized that if they could understand the very small they could also unlock the secrets of the very large, since at the beginning of time everything came from the tiniest of seeds.

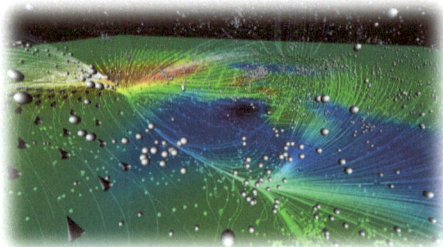

One of the most interesting aspects of string theory is that there may be many more dimensions than we presently see.

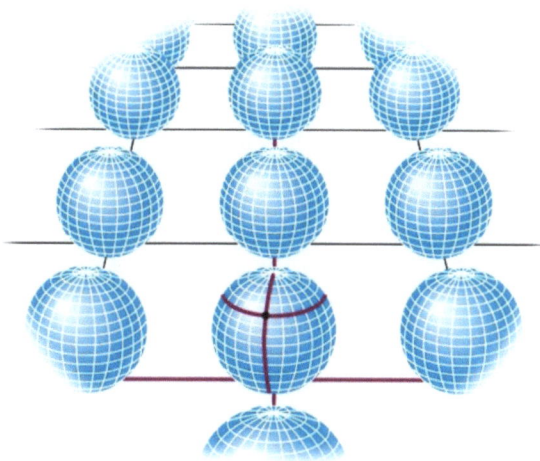

We know of four dimensions: length, width, depth, and time.

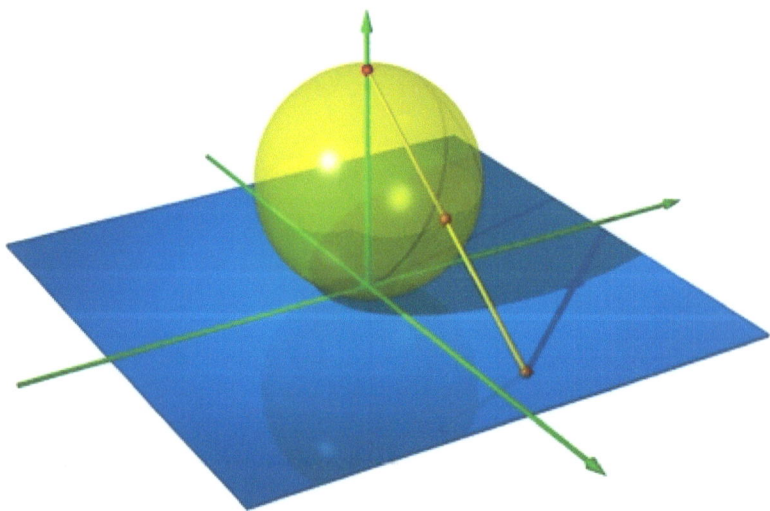

But string theory says there may be 11 dimensions that are curled up so small that we cannot access them.

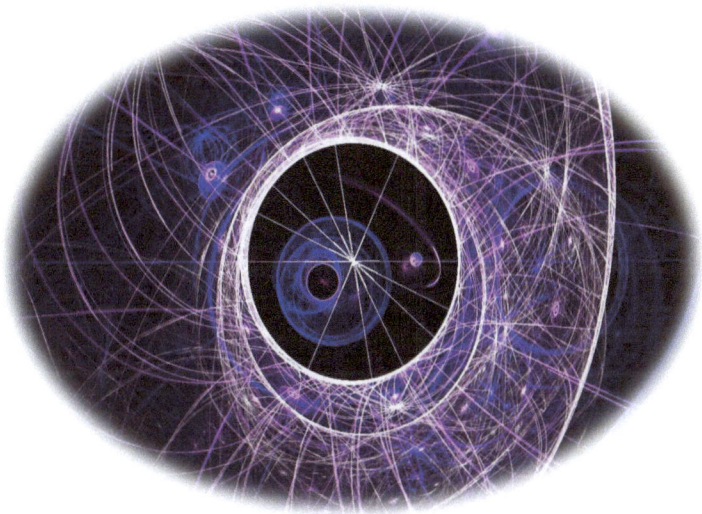

I couldn't visualize this at all, but Brian Greene, the famous Professor of Physics at Columbia University, in a film series gave a good illustration of it.

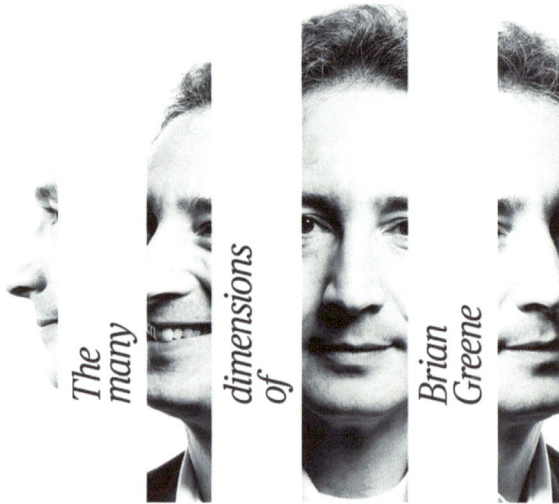

The many dimensions of Brian Greene

He pointed to a lawn and said that if you were an ant the blades of grass would seem like trees, but if you were up in an airplane looking down at that same lawn it would flat and dimensionless.

One of the coolest ideas being suggested by science is that the universe we live in may be only one of trillions of other universes. They have called this the multiverse.

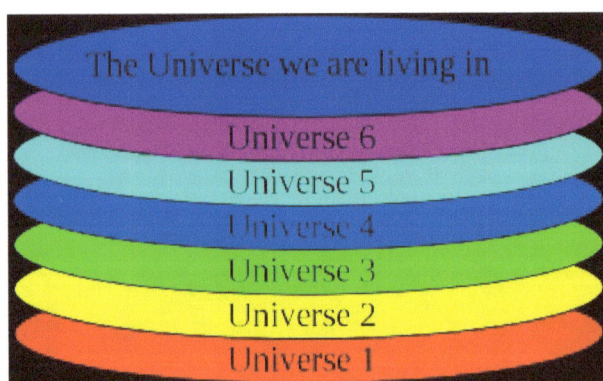

The Universe we are living in

Universe 6

Universe 5

Universe 4

Universe 3

Universe 2

Universe 1

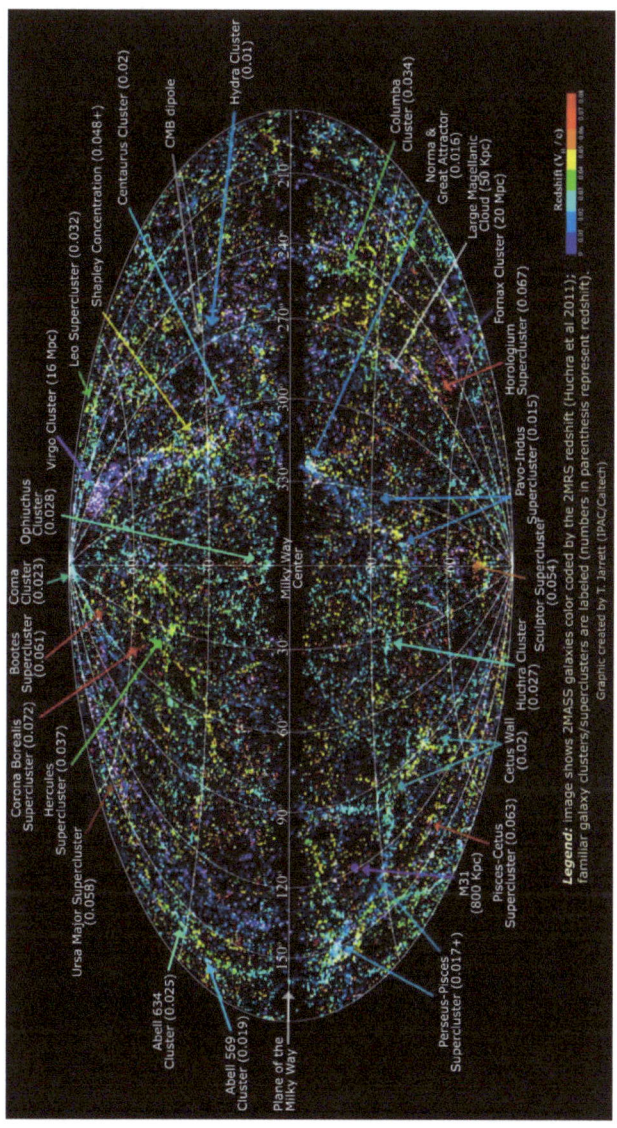

Legend: Image shows 2MASS galaxies color coded by the 2MRS redshift (Huchra et al 2011); familiar galaxy clusters/superclusters are labeled (numbers in parenthesis represent redshift).

Graphic created by T. Jarrett (IPAC/Caltech)

Hydra Cluster (0.01)

Columba Cluster (0.034)

Centaurus Cluster (0.02)

CMB dipole

Norma & Great Attractor (0.016)

Shapley Concentration (0.048+)

Large Magellanic Cloud (50 Kpc)

Leo Supercluster (0.032)

Virgo Cluster (16 Mpc)

Fornax Cluster (0.067)

Ophiuchus Cluster (0.028)

Horologium Supercluster (0.067)

Coma Cluster (0.023)

Milky Way Center

Pavo-Indus Supercluster (0.015)

Bootes Supercluster (0.061)

Sculptor Supercluster (0.054)

Corona Borealis Supercluster (0.072)

Hercules Supercluster (0.037)

Huchra Cluster (0.027)

Cetus Wall (0.02)

Ursa Major Supercluster (0.058)

Pisces-Cetus Supercluster (0.063)

M31 (800 Kpc)

Abell 634 Cluster (0.025)

Perseus-Pisces Supercluster (0.017+)

Abell 569 Cluster (0.019)

Plane of the Milky Way

Redshift (V$_p$ / c)

Apparently, all of these universes may result from slight changes in the vibrations of extraordinarily tiny strings of energy/matter, just as a few chord changes can produce a new song.

The problem with string theory is that it hasn't yet been proven by

The Scientific Method

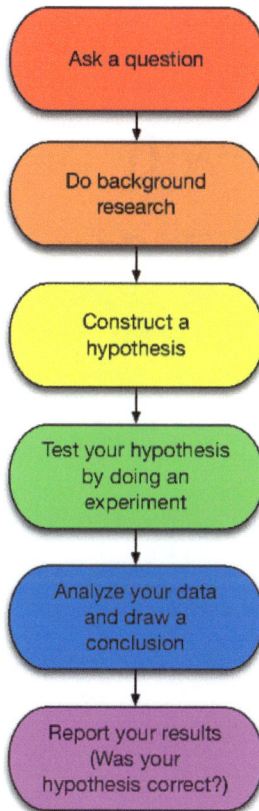

Ask a question

↓

Do background research

↓

Construct a hypothesis

↓

Test your hypothesis by doing an experiment

↓

Analyze your data and draw a conclusion

↓

Report your results (Was your hypothesis correct?)

63

However, scientists have proposed a number of experiments, including some connected to the Large Hadron Collider, which will be able to demonstrate whether string theory is true or not.

Because string theory is open to being tested it is not simply philosophy or wishful thinking.

It doesn't matter how beautiful your theory is, it doesn't matter how smart you are. If it doesn't agree with experiment, it is wrong.

--Richard Feynman, Physicist

I never realized that a simple golf ball when broken down could be so complex when it is viewed from smaller and smaller scales.

My dad suggested that a quote from the poet Tennyson actually summarizes my project perfectly, "if we could but understand a single flower we would know who we are and what the world is."

I found it amusing
when I thought of how
my golf ball was made of
smaller rubber bands
which when squished
up appeared solid.

This got me to thinking about string theory in a different way: maybe the universe, like my golf ball, is really just made of tiny rubber bands that we cannot see!

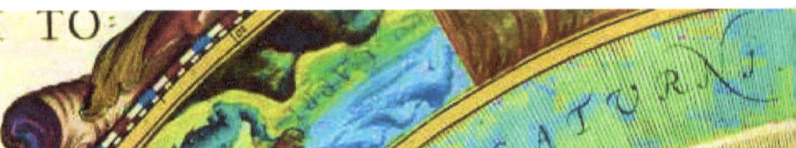

Or, maybe
I am also a unique
strand of strings.

From Strings to Flowers

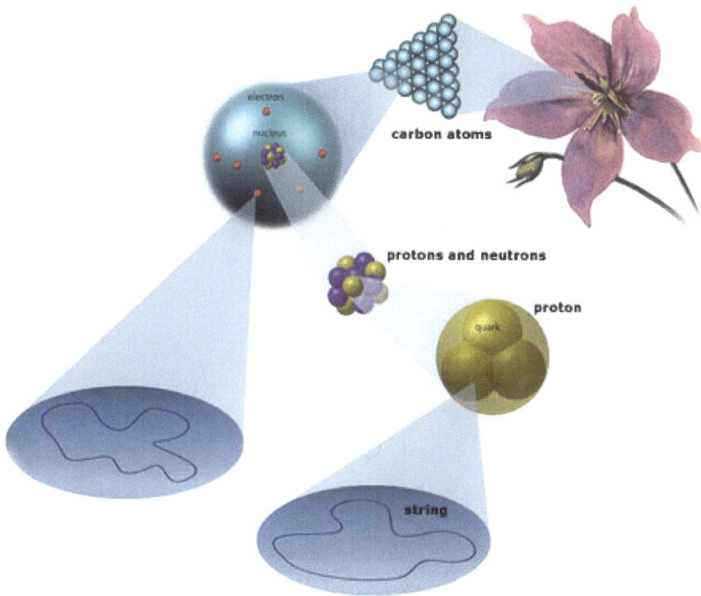

About the Author

Shaun-Michael Lane was born on August 22, 2000. He has one younger brother named Kelly-Joseph. Shaun became fascinated with science at a very early age. His favorite subjects are mathematics, physics, film production, and computer science. His personal hobbies are video games (such as Minecraft), computer programming, and surfing.